Nelson Honiasse Pulaze Luís
Sérgio Alfredo Bila
Arão Raimundo Finiasse

# Diversity and distribution of mammals in Pomene National Reserve

Nelson Honiasse Pulaze Luís
Sérgio Alfredo Bila
Arão Raimundo Finiasse

# Diversity and distribution of mammals in Pomene National Reserve

## Wildlife Management and Protection

**Imprint**

Any brand names and product names mentioned in this book are subject to trademark, brand or patent protection and are trademarks or registered trademarks of their respective holders. The use of brand names, product names, common names, trade names, product descriptions etc. even without a particular marking in this work is in no way to be construed to mean that such names may be regarded as unrestricted in respect of trademark and brand protection legislation and could thus be used by anyone.

Cover image: www.ingimage.com

This book is a translation from the original published under ISBN 978-620-6-76070-2.

Publisher:
Sciencia Scripts
is a trademark of
Dodo Books Indian Ocean Ltd. and OmniScriptum S.R.L publishing group

120 High Road, East Finchley, London, N2 9ED, United Kingdom
Str. Armeneasca 28/1, office 1, Chisinau MD-2012, Republic of Moldova, Europe
Printed at: see last page
ISBN: 978-620-7-75622-3

# GENERAL CONTENTS

# DEDICATORY

*I dedicate this work to my dear and loving parents, Honiasse Pulaze Luís and Emília Bito Sabonete, to my sister Lúcia Luís and my brother Milton Luís for the love, affection and trust they have placed in me during this long journey.*

# ACKNOWLEDGEMENTS

I thank Almighty God, the creator of heaven and earth, for life. To my parents, Honiasse Pulaze Luís and Emília Bito Sabonete for their support, patience, encouragement, unconditional love, giving and help at all times.To Dr Sérgio Alfredo Bila and Dr Arão Raimundo Finiasse, as tutor and co-tutor, for their teaching, encouragement, support, dedication, patience, availability and for believing in my potential. Thank you very much.To the Higher Polytechnic Institute of Gaza and all the teaching staff, in particular Dr° . Sérgio Alfredo Bila, Dr° . Arão Raimundo Finiasse, Eng$^a$ Juvência Malate, Eng° . Severino Macoô, Eng° . Edson Massingue, Eng° . Emídio Matusse and Eng° Pedro Wate, for having contributed greatly to my training.The Manica Provincial Directorate of Education and Human Development (Scholarship Institute) for the scholarship awarded.To the staff of the Pomene National Reserve in particular, Ivone Covane, Óscar, Odilão Macamo and Manuel for their support in carrying out the fieldwork, who were by my side day after day during data collection. In addition to the above-mentioned employees, I would also like to thank Dr Pedro Pereira (administrator) and Dr Aquiles Nhangave (substitute administrator) at RNP for agreeing to collect the data.To my siblings, Milton, Dickson, Claudino, Alfredo, Flora, Lúcia and Cláudia for their encouragement, trust, love and respect. To my colleagues and course mates, particularly those from the 2014 intake (Araújo Sevene, Iolanda Chibamo, Herménio Nhamangua, Valdimiro Soqueres, Jorge Mabote, Simões Elias, Raimundo Manuel, Raimundo Cossa, Arlindo Monguela, iHelton Santomeia, Valdemiro Neve, Edna Mutambe, Lina Chiziane, Xadreque, Durcênia, Rabeca, Eunice, Terço and "Benito José Macopa-in memory") for their company, debates, friendship and discussions. To my friends, Gulamo Abubacar, Soares, Elcídio Rofino, Zemenino Nhacafula and Débora Luís for their friendship and unconditional love.To all the staff at the Matusse condominium and everyone who helped me, directly or indirectly, and made this work possible.

# SUMMARY

*The study on the Diversity and Distribution of Medium-Large Mammals was carried out in the Pomene National Reserve, with the aim of studying the diversity and distribution of medium-large mammal species in the 3 habitats (Dense Miombo, Open Miombo and Shrub Prairie) of the RNP. To collect data using the direct method, all the observation paths in each habitat were travelled on foot in the morning and at the end of the day. Each walk was 2km long and as soon as a herd was observed within a 1km radius of the observers, the species was identified, the total number of individuals of each species was counted and the geographical coordinate of the location was recorded. Using the indirect method, 16 transects of 1000 metres in length and 100 metres in diameter were randomly established 500 metres apart to collect samples of nocturnal animals. On each transect, species were identified using their traces (faeces and footprints) and the total number of individuals of each species was recorded. Simpson's index was calculated to determine specific diversity, Berger-Parker's index to determine specific dominance, Jaccard's similarity index was used to determine specific similarity between habitats in the Pomene National Reserve and software (Quantum GIS and Google Earth) was used to map the distribution of medium-large mammal species in each habitat in the Pomene National Reserve. Nine species of medium-large mammals were recorded in the RNP, belonging to four families and three orders. Miombo Aberto was the habitat with the highest species diversity, with a Simpson's index of 0.846. The most dominant species in all habitats were the grey goat (Sylvicapra grimmia) and the African wild pig (Potamochoerus porcus), and the least dominant species were the chipene (Raphicerus campestris) and the black-faced monkey (Cercopithecus pygerythrus). The highest specific similarity was observed between Dense Miombo and Shrubby Prairie, with a Jaccard similarity index of 0.875. Open Miombo and Shrubby Prairie were the habitats with the highest occurrence of species and the distribution of species in all habitats was uneven.*

***Keywords****: Diversity, Distribution, Mammals.*

# 1. INTRODUCTION

Species diversity refers to the variety of species of living organisms in a given community, habitat or region. Its concept involves two parameters, specific richness and evenness (Franco, 2013).

The diversity and occurrence of mammals in a given region are influenced by abiotic (temperature, rainfall, humidity and soil) and biotic (competition and predation) factors (Peroni and Hernández, 2011). Environmental temperature, rainfall, soil quality and humidity influence the availability of green leaves on plants and the availability of water for wildlife (Dunnin et. al., 1992). Habitats with a greater availability of forage plants, water and shade have a greater diversity and occurrence of mammals (Peroni and Hernández, 2011). Animals look for habitats with a high availability of forage plants and water to satisfy their metabolic and nutritional requirements and restrict their movements within these habitats (Dunnin et. al., 1992).

The occurrence of mammals in a habitat can vary seasonally, according to the changing seasons (Archie and Chiyo, 2012). During the rainy season, due to the high availability of forage plants, animals are able to live in groups, which move in intact clusters and select forage of high nutritional value, grazing in places that bring together the preferred forage, water and shade (Vence et. al., 2008 cited by Clavete, 2014). During the dry season, the low growth of forage plants and the reduction or absence of water bodies in some habitats causes a progressive decline in the biomass of the forage preferred by the animals and they are forced to move to other habitats that have minimal resources available (food, water and shelter) (Macandza et al., 2004).

Biotic factors (competition and predation) also have a strong influence on the diversity and occurrence of medium-sized mammals in a given habitat (Cassini, 2005). In habitats where there is competition between wildlife species, when competition is intense, some animals are usually excluded from using the habitat and are forced to look for other habitats where there is less competition for resources (Peroni and Hernández, 2011). Predation can influence the occurrence of prey and predator populations in habitats (Korpimaki et. al., 1991 cited by Simões, 2009). In habitats where there is predation, there are prey that abandon these habitats as a way of avoiding attack by predators, but there are prey that, despite facing various risks (attacks by predators), remain in these habitats to exploit the existing resources (Peroni and Hernández, 2011). Habitats that have a low occurrence of prey tend to reduce the population of their predators, as they move to other habitats with favoured prey (Peacor and Werner, 2001). Predators have the potential to regulate the distribution and abundance of their prey in a habitat, and vice versa (Schiesari, 2016).

Some of the consequences of resource limitation provide information on how species adapt to the limits of tolerance to which they are subjected, and there may be competition for limited resources, both between organisms of the same species and also between individuals of different species (Esteves, 2010).

In conservation areas, in addition to abiotic and biotic factors, the diversity and distribution of animals can also be influenced by habitat connectivity (ecological corridors and permeability matrices) and historical events (Esteves, 2010). Mostly habitats with better connectivity (ecological corridors and permeability matrices) and an excellent evolutionary history have greater diversity and occurrence of mammals (Eduardo, 2014).

The study of the diversity and distribution of medium-sized terrestrial mammals in the Pomene National Reserve is of great importance and provides important results for the staff of the Pomene National Reserve to propose management measures for the area in terms of the method to be used for the management of each habitat, based on their behaviour and their dependence on resources, encouraging their conservation.

## 1.1. Study problem and justification

The diversity and abundance of wildlife in the Pomene National Reserve was already quite low at the time it was created, and now over the years it tends to be in a worse state (MITADER, 2016). The main causes for the reduced faunal diversity and abundance are associated with the fact that the Reserve is small, the use of uncontrolled burning, invasion for the construction of human dwellings, transformation into agricultural areas and for cattle grazing (Macandza et. al., 2015).

The invasion of the Reserve for the construction of human dwellings, agricultural practices and cattle grazing reduces the size of the wildlife habitat, and the incidence of uncontrolled fires contributes to the reduction in the availability of forage plants for wildlife (Joia et al., 2016).

Reduced habitat size and availability of forage plants can influence the low growth and distribution of wildlife (Joia et. al., 2016). Wildlife needs space to carry out their daily activities and sufficient quantities of forage to fulfil their metabolic needs (Muchanga et. al., 2016). The low availability of forage in some habitats can influence wildlife travelling in search of habitats with a high availability of food, water, shade and shelter (Dunnin et. al., 1992). Travelling long distances in search of habitats with good conditions can cause species to lose energy, which would be important for their reproductive activities. The occupation of new habitats with the availability of food, water and shelter can restrict the distribution of animals in these habitats rather than in their usual habitats and this can hinder the knowledge of wildlife occurrence areas and consequently the planning of wildlife management activities (Muchanga et. al., 2016). The lack of management activities can jeopardise the growth of wildlife populations, as well as the conservation of fauna species in the Pomene National Reserve (Muchanga et. al., 2016). This study is extremely important because it will provide maps of the occurrence of mammal species, information on the species that exist in each habitat and their dominance, with the aim of facilitating the implementation of wildlife management activities in the Pomene National Reserve.

## 1.2. Objectives 1.2.1.General

➢ To study the diversity and distribution of medium-large mammals in the Pomene National Reserve.

### 1.2.2. Specific

➢ To determine the diversity, dominance and specific abundance of medium-large mammals in each habitat of the Pomene National Reserve;

➢ To determine the specific similarity of medium-large mammals between habitats in the Pomene National Reserve;

➢ Map the distribution of medium-large mammals in each habitat of the Pomene National Reserve.

## 1.3. Study hypotheses

**H1:** The dense miombo has a greater diversity and occurrence of medium-large mammals in the Pomene National Reserve.

**H2:** The shrub grasslands have a lower diversity and occurrence of medium-large mammals in the Pomene National Reserve.

<h1 align="center">2. LITERATURE REVIEW</h1>

**Mammals**

Mammals (from the Latin Mammalia) are a class of vertebrate animals in the domain Eukaryota, of the Kingdom Animalia and Phylum Chordata, subdivided into two groups: aquatic (cetaceans) and terrestrial (quadrupeds/bipeds), which are characterised by the presence of mammary glands which, in females, produce milk to feed the young (or calves), the presence of hair or fur, with the exception of dolphins and some whales, which only in the embryonic stage have hair (Mundo Educação, 2014). According to their body mass, mammals can be considered small (under 10kg), medium (between 10kg and 50kg) and large (over 50kg) (Bernardo, 2012).

## 2.1. Mammal diversity in Mozambique

Mozambique is home to 271 mammal species (Pereira and Nazerali, 2016), of which 104 are considered threatened and protected species, 21 endemic or near-endemic species and 3 introduced species (Schneider et. al. 2005). Mammal diversity is highest in Sofala and Manica Provinces, particularly along the Chimanimani mountains and in the east of Gaza Province (Pereira and Nazerali, 2016). Endemic mammal species are concentrated in the Serra da Gorongosa - Rift Valley - Marromeu Complex region, in the rugged Chimanimani region, along the coast of the southern provinces, particularly between Vilanculos and Inhambane and south of the Save River, near Inhassoro and Zinave (Schneider et. al., 2005).

## 2.2. Distribution of mammals in Mozambique

The distribution of threatened and protected mammals is similar, but fewer species occur along the coast of Inhambane, Gaza and Maputo Provinces

(Pereira and Nazerali, 2016). Particularly in the Bazaruto and Inhambane Archipelagos, there are several endemic species of mammals, birds and reptiles, probably due to the isolation of the islands (Schneider et. al., 2005). In Cabo-Delgado, the highest densities of animals are found in the east, near the coast, and the highest concentration of mammals is in the north-east of the province (SEED, 2003).

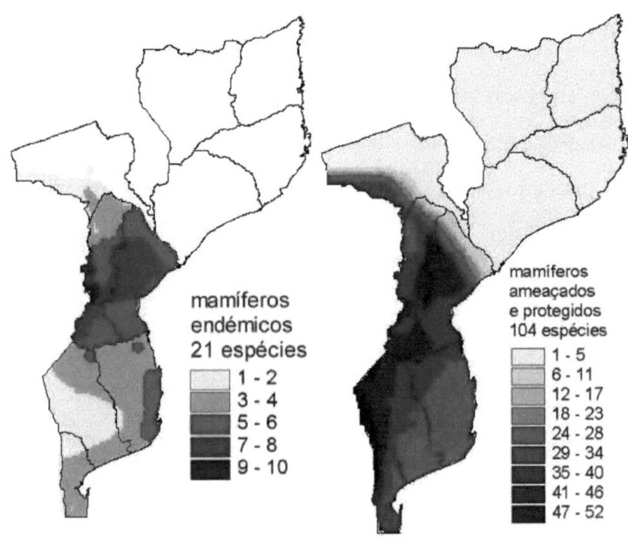



**Source:** Schneider et. al., 2005.

**Map 1:** Distribution of mammals in Mozambique.

## 2.3. Diversity scales

Three types of diversity can be distinguished: alpha ($\alpha$) or local diversity, which corresponds to diversity within a habitat or community, beta ($\beta$) diversity, which corresponds to diversity between habitats or any other environmental variation, measuring how much species composition varies from one place to another, and gamma ($\gamma$) or regional diversity, corresponding to the diversity of a large area, biome, continent, island, etc (Barros, 2007).

## 2.4. Measures of diversity. 2.4.1.Shannon-Wiener indices (H')

It was proposed by Shannon in 1948 and has an advantage over the Margalef, Gleason and Menhinick indices (Rodrigues, 2015). The most widely used diversity index is the Shannon-Wiener index (H'), which gives greater weight to rare species (Barros, 2007).

$$\text{Índices de Shannon–Wiener } (H') = -\sum_{i=1}^{s} \frac{ni}{Ni} \ln\left(\frac{ni}{Ni}\right)$$

Equation [1]

Where:

ln = Neperian base logarithm (e); ni = number of individuals sampled for speciesi;
N = total number of individuals sampled.


### 2.4.2. Simpson's index (Ds).


This index was proposed by Simpson in 1949 and has an advantage over the Margalef, Gleason and Menhinick indices, as it not only considers the number of species (s) and the total number of individuals (N), but also the proportion of the total occurrence of each species (Rodrigues, 2015). It expresses the probability of two randomly selected individuals being of the same species. It ranges from 0 to 1 and the higher it is, the more likely the individuals are to be of the same species (Uramoto et. al., 2005, cited by Biondi and Bobrowski, 2014). Simpson's diversity is estimated using the following equation (Rodrigues, 2015):

$$\text{Simpson's Diversity Index } (\textbf{Ds}) = 1 - \frac{\sum_{i=1}^{n} ni\,(ni-1)}{N(N-1)}$$

Equation [2]

Where: ni is the number$^{\circ}$ of individuals of each species; N is the total number$^{\circ}$ of individuals in each sample.

### 2.4.3. Margalef Index (DMg)

It was proposed by Margalef in 1951 and is a simple diversity index (Rodrigues, 2015). It expresses the species richness of a sample, considering the number of species (S-1) and the logarithm (base 10 or natural) of the total number of individuals (lnN) (Maitini and Prado, 2010). Low species richness is considered to be values less than 2.0 and high species richness is considered to be values greater than 5.0 (Richter et al., 2012) and is estimated using the following equation (Maitini and Prado, 2010):

Index Index Index (DMg)

$$\frac{S-1}{\ln N}$$

Equation [3]

Where: s = n° of all species; N = n° of individuals of all species; ln = natural limit.

### 2.4.4. Berger-Parker Index.

This dominance index was proposed by BERGER & PARKER in 1970. It is a simple index when compared to Simpson's dominance index, but efficient. It considers the greatest proportion of the species with the greatest number of individuals (Rodrigues, 2015).

Berger-Parker Index $(d) = \frac{Nmax}{NT}$ Equation [4]

Where: Nmax is the n° of individuals of the species; NT is the total n° of individuals in the sample.

## 2.4.5. Jaccard's similarity index

The Jaccard Similarity Index (Jaccard Coefficient) takes into account the relationship between the number of common species and the total number of species found (Silvestre, 2009). The Jaccard Index varies from 0 to 1 and the higher the value, the greater the similarity between the habitats (Gomes &

Ferreira, 2004). The Jaccard Index is commonly used for qualitative comparisons of communities, rarely reaching values above 60% (Mantovani, 1987 cited by Ferraz, 2008). Areas with values above 25% of the Jaccard similarity index are considered similar (Mueller-Dombois and Ellenberg 1974). One of the great advantages of this index is its simplicity (Gomes & Ferreira, 2004 cited by Roque, 2015).

$$\text{Indice de Similaridade de Jaccard } (ISJ) = \frac{a}{a + b + c}$$

Equation [5]

Where: a = Number of common species in habitat A and B;

b = Number of species that occur only in habitat A;

c = Number of species that occur only in habitat B.

## 2.5. Distribution

The distribution of animals can be defined as the occurrence and spatial arrangement in an area occupied by the population or species (Esteves, 2010), it is the way in which animals are dispersed in a given area or habitat (Begon, 2006).There are three basic patterns of spatial distribution of individuals (Brower and Zar, 1984), random distribution, which is mainly observed in homogeneous environments, which allows individuals to be randomly spaced apart, regular or uniform distribution, which is perceived when competition between individuals is severe, promoting spacing by a constant minimum distance between individuals, and aggregate distribution (contagious or clustered), which is the most common pattern (Odum and Barrett, 2008).

## 2.6. Methods used to study the diversity and distribution of mammals.

### 2.6.1. Observation method indirect.

This method consists of marking out squares or transects in each habitat, and species can be identified using traces (footprints and faeces), other indirect traces that can provide satisfactory data for the study are recorded, such as torn branches and the type of bite on the species consumed (Neto et. al., 1995). Sampling for traces in squares or transects should be carried out during dry periods (months of lower rainfall), as these are the months when animals are most active in general. This sampling requires a great deal of technical and practical knowledge from the field team, since most of the time, no matter how good the soil and climate characteristics are, the infinite conditions of free-living animals mean that the prints left behind are quite different from the standard average shapes and measurements of the species (Cunha, 2013).

### 2.6.2. Direct observation method.

It consists of observing the animal in real time at the site, recording data on the species observed and the coordinates of the places where the species are found in each sample (Roos, 2010).

Studying the diversity and distribution of mammals using the direct method can be done using camera traps, driving along carriageways and walking along existing trails (Bothma, 2002). The species are identified and the total number of individuals of each species observed is recorded, followed by travelling to the observation site to record the geographical coordinates (Roos, 2010).

**2.6.2.1. Techniques applied to the direct observation method for the study of wildlife.**

**2.6.2.1.1. Setting up photographic equipment (camera traps).** The trap (camera trap) is installed in locations chosen during the walks, taking as a parameter the characteristics of the vegetation (presence of fruiting specimens and shading), presence of tracks or other traces, in places close to watercourses and preferably in places of possible passage according to prior registration and identification of travel points (Araujo and Chiarello, 2007). Readings should be taken daily, at regular intervals for general maintenance (renewing the battery and batteries when necessary, cleaning and checking the state of operation), and fixed to trees with a diameter of more than 15cm at a height of approximately 45cm from the ground (Marques and Ramos, 2001).

**2.6.2.1.2. Carriageways**

It consists of searching for species by travelling in a car at low speeds, usually less than 40 km/h along main roads and access routes between habitats (Bothma, 2002). After observing the herd, the species are identified and the total number of individuals of each species observed is recorded, followed by travelling to the observation site to record the geographical coordinates (Roos, 2010).

**2.6.2.1.3. Observation paths**

The study is conducted within the area, slowly on foot looking for species in all visually accessible habitats and in places close to watercourses (Bothma, 2002). Walks are carried out in the early morning (before one hour after sunrise) and late afternoon on existing trails, establishing a radius of 1 to 2 kilometres from the observers (Roos, 2010). They can stop every 100 metres to allow the animals to calm down and the observers the chance to listen carefully to the sounds and watch the animals (Bothma, 2002).

## 2.7. Geographic Information Systems (GIS)

Geographic information systems (GIS) handle geographically referenced data as well as non-spatial data, including operations to support spatial analyses (Maguire et. al., 1991). Geographic Information Systems can be defined as computer systems capable of capturing, storing, querying, manipulating, analysing and printing data spatially referenced to the earth's surface (Filho and Iochpe, 1996).

### 2.7.1. Google Earth

Google Earth (GE) is free software that combines satellite images with terrain features to provide a 3D digital rendering of the Earth's surface in an interface that is considered easy to manipulate for the end user and has vast application potential for both the corporate world and for academic purposes (Lima, 2012). The set of tools currently provided by Google Earth offers resources for mapping, importing and exporting GIS (Geographic Information System) data and detailed 3D visualisation of practically the entire surface of the planet using satellite images and high-resolution historical aerial photos (Filho and Iochpe, 1996).

### 2.7.2. Global Positioning System (GPS)

The Global Positioning System (GPS) is a navigation system based on satellite signals, made up of a network of 24 satellites placed in orbit by the US Department of Defence (Gomes, 2010). The GPS system is made up of three segments, the space segment (artificial satellites on Earth that emit electromagnetic signals), the control segment (ground stations that keep the satellites in operation) and the user segment, made up of receivers that pick up the signals sent by the satellites in order to calculate their position (Bernardi and Landim, 2002). GPS technology is essential in a number of areas, particularly mapping, military, maritime, air and land navigation (Rodrigues and Reis, 2014). The great advantage of this system is its ability to integrate with other systems, such as the Geographic Information System (GIS), capable of producing digital maps in real time with high precision (Maguire et. al., 1991). GPS is the key to bringing these two together systems, as it initially allows for the acquisition of data, which will form the geometric basis for spatial analysis by GIS (Gomes, 2010).

### 2.7.3. Importing and exporting GIS data

The data generated on the different GIS platforms can be exported for visualisation in Google Earth in two ways. In the professional version (Google Earth Pro) it is possible to directly import a range of GIS data formats and in the free version it is necessary to convert the files to KML format using other tools and programmes. GIS software already has tools for exporting data in KML format for use in Google Earth (Lima, 2012).

### 2.7.4. Mapping

The cartographic field was one of the first to take advantage of the GPS system because it led to a drastic increase in accuracy, making it easier to read and build maps. Any organisation or agency can benefit from the efficiency and productivity of GPS devices (Rodrigues and Reis, 2014). Google Earth has tools for editing vectors in point, line and polygon formats, allowing for the mapping of features and cartographic representation of elements identified through satellite images (Lima, 2012).

### 2.8. Factors influencing the distribution of mammals.

The distribution of species can be seen as a spatial reflection of their niche; species occur where environmental conditions are favourable and where one or more essential resources are present (Esteves, 2010). In habitat complementation, species move between different types of environments to obtain non-substitutable resources (Dunnin et. al., 1992). An environment in which species or populations live has a series of interactions complex, between biotic and abiotic factors, which influence their distribution and abundance (Esteves, 2010).

## 2.8.1. Abiotic factors.

These are caused by climate, temperature variation, the presence or absence of adequate light and humidity (Gehring and Swihart, 2003). Each species can only live within a certain tolerance range of these factors; below these limits their vital functions are seriously compromised and they die (Peroni and Hernández, 2011). However, even within the tolerance limit, for each species there is a suitable level at which its development is maximised (Esteves, 2010). The control of internal temperature gives homeothermic animals a great advantage: regardless of variations in the environment, their bodies will maintain a constant temperature, while the influence of temperature on poikilothermic animals will be felt more strongly (Gehring and Swihart, 2003). The most important condition for the life of organisms is temperature, and it can act at any stage of the life cycle and limit the distribution of a species through its effects on survival, reproduction, growth and interaction with other life forms (Peroni and Hernández, 2011).Humidity can also determine the distribution limits of some species, depending on altitude, and resistance to drought is an important ecological characteristic that varies from one species to another (Esteves, 2010). The role of humidity is more evident on a global scale, as the detailed way it acts on species on local scales is not always clear (Peroni and Hernández, 2011). Light can limit the distribution of species by influencing animal behaviour, since photoperiod influences some phenomena such as reproduction and the rate of migration (Esteves, 2010).

## 2.8.2. Biotic factors

This is due to the availability of resources and habitat characteristics, such as the shape, size, connectivity and permeability of the matrix (Esteves, 2010). It is known that mammal species perceive the landscape from different perspectives, with large and generalist species exhibiting a relatively small difference in terms of resource selection (Gehring and Swihart, 2003).

The movement of animals between landscapes is a common and necessary process, as it allows organisms to establish their territories and living areas, and this movement depends on the spatial attributes of the landscape in which they live (Peroni and Hernández, 2011). Characteristics such as the density, size, degree of aggregation and connectivity of habitats tend to control colonisation rates and the risk of extinction (Esteves 2010).

In places where species or populations live, there are a series of intraspecific or homotypic interactions, those that occur within the same species, and interspecific or heterotypic interactions, when they occur between individuals of different species (mutualism, parasitism, competition and commensalism) (Cassini, 2005).Some consequences of resource limitation are very important for understanding how species adapt to the limits of tolerance to which they are subjected. There can be competition for limited resources, both between organisms of the same species and also between individuals of different species (Esteves, 2010). Competition is an interaction between individuals, caused by a common need for a resource and leading to a reduction in the survival, growth and reproduction of at least some of the competing individuals involved (Peroni and Hernández, 2011).After species are introduced into a given area, they tend to move to places with minimal resources and favourable conditions for their survival, affecting the migration rates of animals between habitats and possibly causing the extinction of some species in places without the desired conditions (Esteves, 2010).

## 2.9. Sampling

Sampling can be defined as the mechanism applied to obtain information (collect data) from a small part (sample) of a large group (population) and learn something (conclude) about this larger group (population) (Henriques, 2012). The sample is obtained from a well-defined population using processes that are well-defined by the researcher and is subdivided into two groups (probabilistic and non-probabilistic) (Barbetta, 2002). Sampling is probabilistic if all the elements of the population have a known, non-zero probability of belonging to the sample, otherwise sampling is non-probabilistic (Henriques, 2012).

### 2.9.1. Sampling probabilistic.

The probability sampling method requires that each element of the population has the same probability of being selected (Cunha, 2017). Probabilistic sampling involves a draw with well-defined rules, which can only be carried out if the population is finite and fully accessible (Henriques, 2012).

### 2.9.1.1. Simple Random Sampling

All members of a population have the same probability of being included in the sample and it is suitable for homogeneous populations (Henriques, 2012).

### 2.9.1.2. Systematic Sampling

The population must be ordered in such a way that the elements are identified by position, which is also indicated for homogeneous populations (Barbetta, 2002).

### 2.9.1.3. Stratified Sampling

It consists of dividing the population into more homogeneous subgroups (strata), so that there is homogeneity within the strata and heterogeneity between the strata (Henriques, 2012).

**Uniform** - An equal number of elements are drawn from each stratum (Henriques, 2012).

**Proportional** - The number of elements in each stratum is proportional to the number of elements in the stratum (Henriques, 2012).

**Optimal** - A number of elements are taken in each stratum that is proportional to the number of elements in the stratum and also the variation of the variable of interest in the stratum, measured by its standard deviation (Henriques, 2012).

### 2.9.1.4. Conglomerate Sampling

This sampling scheme is used when there is a subdivision of the population into groups that are quite similar to each other, but with strong discrepancies within the groups, so that each one can be a small representation of the population of specific interest (Vieira and Bessegato, 2013).

### 2.9.2. Non-probabilistic sampling.

These are samples in which there is a deliberate choice of sample elements (Andrade, 2017).

### 2.9.2.1. Accidental Sampling

The elements of the sample are chosen accidentally (Andrade, 2017).

### 2.9.2.2. Intentional Sampling

The researcher intentionally targets a group of elements from which they wish to collect information (Andrade, 2017).

### 2.9.2.3. Quota Sampling

It is widely used in market research and political opinion polling where time and money are scarce (Andrade, 2017).

# 3. METHODOLOGY

## 3.1. Location of the study area

The Pomene National Reserve is located in the south of Mozambique, in the province of Inhambane, in the district of Massinga, situated in the centre of the province of Inhambane, with the districts of Funhalouro and Morrumbene to the south, the district of Funhalouro to the west, the district of Vilankulo to the north and north-east, and the Indian Ocean to the east (MAE, 2014).

**Map 2:** Location of the Pomene National Reserve

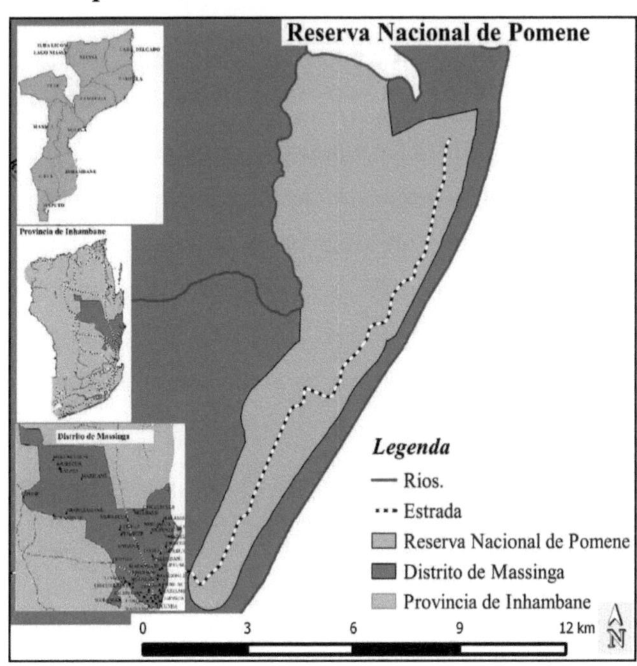


**Source:** Author, 2018.

## 3.2. Topography and Soils

The RNP is made up of low areas (0 to 25 m high), generally the topography of the Reserve is characterised by higher altitudes along the coastal dunes, up to 125 m, decreasing towards the west as we approach the lower Muducha River and from south to north decreasing towards Pomene Bay (INIA, 1995). Most of the RNP is made up of sandy dune soils, along the lower Muducha River to Pomene Bay they are sandy hydromorphic and along the dunes the soil is typical of yellowish coastal dunes (MITADER, 2016).

## 3.3. Weather

The climate in the RNP region is tropical dry and humid with average annual rainfall between 650 and 750 mm and an average annual temperature of $22.9°$ C (Macandza et. al., 2015). The RNP has two distinct seasons, the hot and rainy season which runs from November to April, with January being the hottest month of the year with an average temperature of $28.6°$ C, and the cool or dry season, from May to October, with July being the coldest month of the year, with an average temperature of $19.0°$ C (MICOA, 2013). The average annual rainfall in the rainy season is 1,200 mm, with the highest incidence in February and March (MAE, 2014).

## 3.4. Drainage and Hydrology

Two small rivers run through the RNP: the Muducha River, which forms the western boundary of the Reserve, and the Pedras River, located near the south-eastern boundary, which flows into the Indian Ocean. The Muducha River flows into Pomene Bay, which is an estuary of great importance in terms of biodiversity and has extensive patches of mangrove (PEA, 2016).

## 3.5. Fauna

The diversity and abundance of wildlife in the RNP was already quite low at the time it was created, and now over the years it tends to be in a worse state (PEA, 2016).In the Pomene National Reserve there are no mammal species of international conservation concern (IUCN Red List), there are only eight species protected by law in Mozambique (PEA, 2016). The distribution of 24 mammal species covers the RNP region (Smithers and Tello, 1976 cited by PEA, 2016), and the presence of the squirrel (Sciurus sp), macaque (Papio ursinus), grey goat (Sylvicapra grimmia), wild pig (Potamochoerus porcus), changane (Neotragus moschatus) and hare (Lepus saxatilis) was confirmed (Macandza et. al., 2015).

### 3.5.1. Threats

Threats to wildlife are associated with anthropogenic activities within the Reserve, which include habitat degradation and reduction, poaching and the use of prohibited means to hunt small species within the Reserve, the transmission of infectious diseases from cattle and goats within the Reserve's boundaries and human-wildlife conflict (PEA, 2016).

### 3.6.  Vegetation and Habitats

According to the vegetation mapping of the Zambezi Flora, the RNP is mainly covered by coastal brine and recent dune forests (Wild and Barbosa, 1967). Six main vegetation types have been identified, the miombo (dense and open), shrub grassland, dune vegetation, riverine vegetation, temporarily flooded grassland and the Mangal (Macandza et. al., 2015).

### 3.6.1. Miombo (dense and open)

This is the most predominant vegetation type in the RNP with an extension of 1,937 hectares (38% of the Reserve), the most dominant species is red messassa (Julbernardia globiflora), followed by messassa (Brachystegia speciformis) and chanfuta (Afzelia quanzensis) (Macandza et. al., 2015). The dense miombo (around 50 per cent of the canopy cover) has a poorly developed grass layer (average biomass of 756 kg/ha), the open miombo (around 20 per cent of the canopy cover) has more scattered trees, with the most abundant grass layer (around 30 per cent of the soil) and is dominated by spear grass (Heteropogon contortus), common finger grass (Digitaria eriantha) and cotton grass (Imperata cylindrica) (PEA, 2016).

### 3.6.2. Shrub prairie

It is the second most prevalent type of vegetation in the RNP, with an area of

1,493 hectares (29.5% of the Reserve), the dominant shrub species are Umnhonsi (Salacia cfcraussi), wild palm (Hyphaene cariacea) and mutumbi (Garcinia livingstonei) (Macandza et. al., 2015). This type of vegetation has a grass cover of 40 per cent and an average biomass of 3,100 kg/ha, which is dominated by the species described for miombo (PEA, 2016).

### 3.6.3. Dune vegetation

It covers 856 ha (17% of the RNP) and is practically natural, except for small areas where you'll find the lighthouse, holiday homes and fishermen's camps (PEA, 2016).

### 3.6.4. Riverside vegetation

It occurs along watercourses and is made up of tall grasses dominated by reeds (Phragmites mauritianus and Coix lacryma) followed by various species of the Cyperus genus (PEA, 2016).

### 3.7. Demographics

Until 1995 there were only 3 families living within the RNP, the year in which immigration into the RNP and outlying areas began (Macandza et. al., 2015). The movement was driven by the availability of natural resources and extensive unoccupied areas, reaching 535 households by November 2014 (PEA, 2016).

### 3.7.1. Economic activities

The agriculture practised within the Pomene area does not differ from the national trend for most rural areas, being rudimentary and geared towards household subsistence (PEA, 2016). This activity, within the Reserve area, is practised not only by residents inside the Reserve but also by households from surrounding areas, due to the scarcity of land outside the Reserve (Macandza et. al., 2015).Animal husbandry within the Reserve is practised on a small scale, concentrating on cattle, goats, pigs and poultry, mostly in the southern region of the Reserve (PEA, 2016). Fishing is practised inside the Reserve, along the Muducha River, in the estuary and along the coastal zone of the Reserve Peninsula, and is the second most practised activity after agriculture (Macandza et. al., 2015).

## 3.8. Materials and Methods

### 3.8.1. Materials:

➢ GPS device;

➢ Logbook;

➢ Tape measure;

➢ Biros;

➢ Guidebook for mammals;

➢ Camera;

➢ Software (Google Earth and Quantum GIS).

### 3.8.2. Methods

The study on the diversity and distribution of medium-large mammals was carried out in Pomene National Reserve. Sampling was carried out in 3 habitats (Dense Miombo, Open Miombo and Shrub Prairie) over 45 days. Direct and indirect observation were used simultaneously to collect the data.

### 3.8.2.1. Data collection using the direct observation method

To collect data using the direct method, all the observation paths in each habitat were travelled on foot. The observation walks were carried out in the morning (from 06:30 to 10:00) and at the end of the day (from 15:00 to 18:30). Each walk covered a distance of 2km, and every 100m was stopped to give the animals a chance to calm down and for the observers to watch the animals. During the walks, as soon as the animals were spotted within a radius (distance) of 1 kilometre from the observer, the species were identified, the total number of individuals of each species was recorded and then the site was moved to record the geographical coordinates. At the end of each observation walk, a distance of 200 metres was walked to start the next observation walk.

### 3.8.2.2. Data collection using the indirect method

To give us the chance to sample animals with nocturnal habits, 48 transects were randomly marked out in 3 habitats, 16 transects in each habitat, 1000m (1km) long and 100m wide, 500m apart. Each transect was used to identify the species and record the total number of individuals of each species using the presence of their traces (faeces and footprints) and then the geographical coordinates of the traces found on each transect were recorded.

### 3.8.3.   Determination of parameters.

### 3.7.3.1.Diversity

To determine the specific diversity in each habitat, Simpson's diversity index was calculated according to equation 1.

### 3.8.3.2 Dominance & Abundance

The Berger-Parker dominance index was used to determine the dominance of each species and the specific abundance in each habitat, according to equation 3.

### 3.8.3.5 Similarity index

Jaccard's similarity index was used to determine species similarity between the 3 habitats studied, according to equation 4.

### 3.8.4. Distribution

To map the vegetation and the distribution of species, the geographical coordinates were organised in a Microsoft Office Excel spreadsheet for each habitat. Once organised, they were exported from Microsoft Office Excel and imported into Google Earth, where they were converted and saved in (kml) format and imported into Quantum GIS in order to produce maps with a detailed visualisation of each habitat (Lima, 2012).

### 3.8.5. Analysing data

### 3.8.5.1. Diversity, Dominance, Abundance and Specific Similarity.

This was done on the basis of comparisons of the results obtained in determining the indices determined for each parameter.

### 3.8.5.2. Distribution.

Data on specific distribution was analysed by mapping the species found in each habitat and then comparing the distribution between habitats.

# 4. RESULTS

## 4.1. Diversity of medium-large mammals in Pomene National Reserve

During the study, 9 species of medium-large mammals were identified, belonging to four families and three orders. The bovidae family had the highest number of species (see Annex 2). The open miombo was the habitat with the highest specific diversity of medium-large mammals, the results indicate that this habitat had a Simpson's diversity index of 0.846, and the shrub grassland with a Simpson's diversity index of 0.805, was the habitat with the lowest specific diversity of medium-large mammals in the Pomene National Reserve (see graph 1):

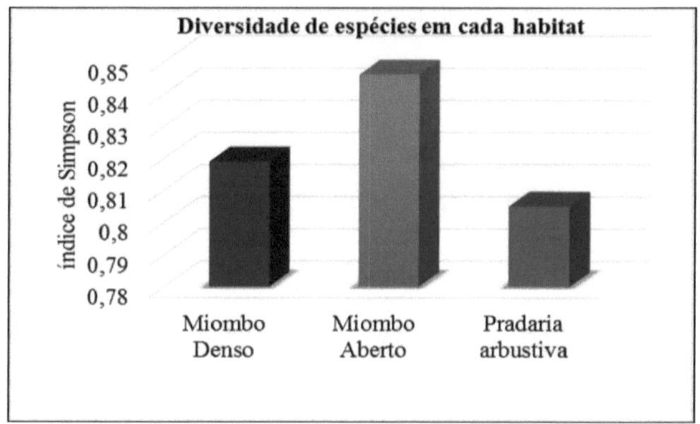


**Graph 1:** Diversity of medium-large mammals in RNP habitats

## 4.2. Dominance of medium-large mammals in RNP habitats.

The results obtained in this study show that the grey goat (Sylvicapra grimmia) was the most dominant species in the dense miombo and open miombo, with a relative dominance of 24.8927% and 20.362% respectively. In the shrub grassland, the most dominant species was the wild pig (Potamochoerus porcus), with a relative dominance of 27.6 per cent. The xipene (Raphicerus campestris) was the least dominant species in the open miombo and shrub prairie, with a relative dominance of 0.905% and 2% respectively. The black-faced monkey (Cercopithecus aethiops) was the least dominant species in the dense miombo, with a relative dominance of 3.8627% (see graph 2).

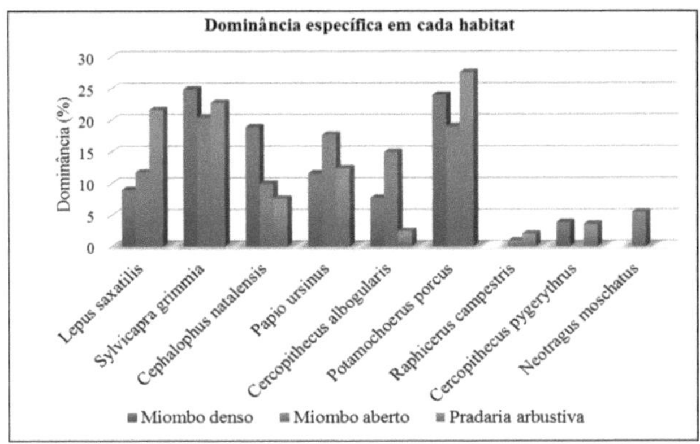


**Graph 2:** Specific dominance of medium-large mammals in RNP habitats.

### 4.3. Abundance of medium-large mammals in the Pomene National Reserve

This study shows that the shrub grassland was the habitat with the highest abundance of medium-large mammals, with a relative abundance of 35.51 per cent, and the habitat with the lowest abundance of medium-large mammals was the open miombo with a relative abundance of around 31.39 per cent. The dense miombo had a relative abundance of 33.10 per cent, as shown in the following graph (graph 4):

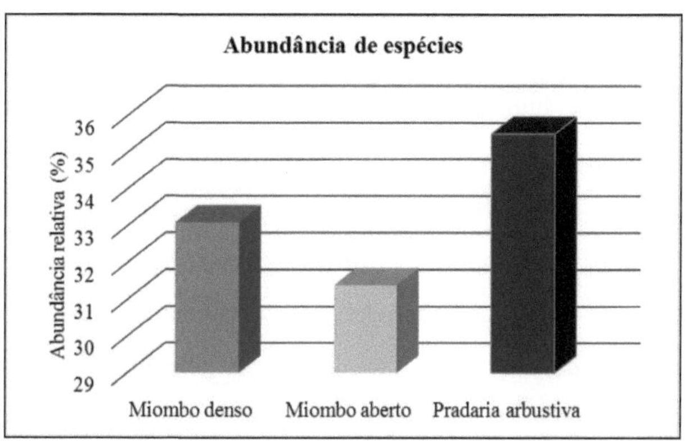


**Graph 3:** Abundance of medium-large mammals in Pomene National Reserve.

## 4.5 Similarity of medium-large mammals between habitats in the RNP

The graph below illustrates the specific similarity of medium-large mammals between habitats in the Pomene National Reserve. The dense miombo and the shrub grassland were the habitats with the highest specific similarity, with a Jaccard similarity index of 0.875. There was less specific similarity between dense miombo and open miombo with a jaccard similarity index of around 0.667 (see graph 5).

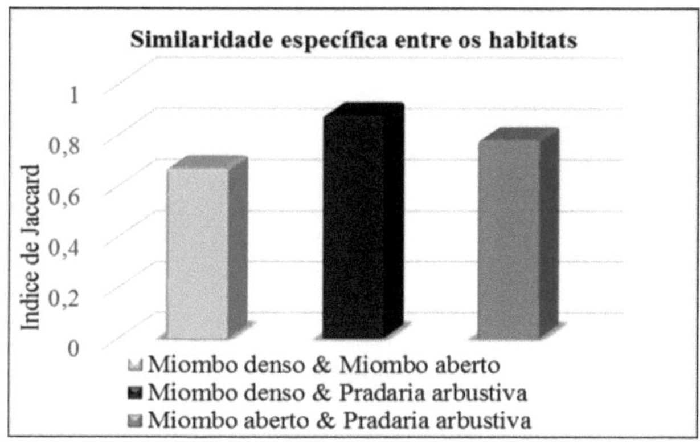


**Graph 4:** Similarity of medium-large mammals between habitats in the RNP.

## 4.6 Distribution of medium-large mammals in the Pomene National Reserve

The habitats with the highest occurrence of animals were open miombo and shrub prairie. Eight species were confirmed in the open miombo and seven in the dense miombo (see map 3 below). The chengane goat (Neotragus moschatus) and xipene (Raphicerus campestris) are the species that were not found in the dense miombo. The black-faced monkey (Cercopithecus pygerythrus) does not occur in the open miombo and the chengane goat in the shrub grassland (Neotragus moschatus). The grey goat (Sylvicapra grimmia), African wild pig (Potamochoerus porcus), and golden-necked hare (Lepus saxatilis) are the species that occur most frequently in all the habitats studied, as illustrated in the map below:

**Map 3:** Distribution of medium-large mammals in Pomene National Reserve.

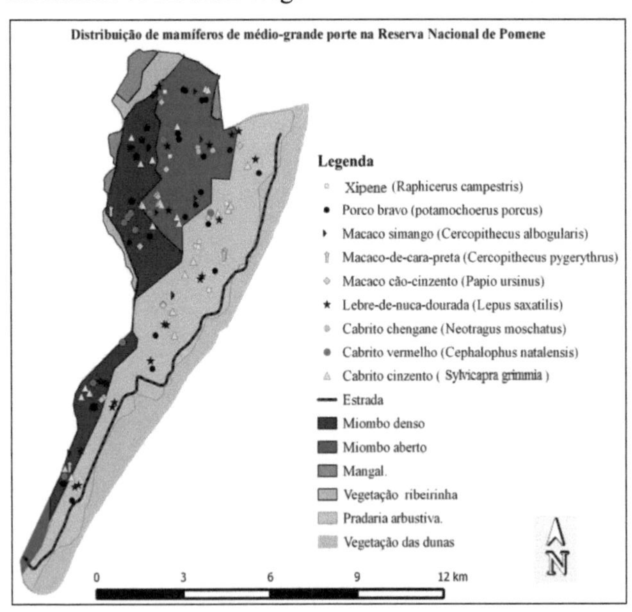


**Source:** Author, 2018.

# 5. DISCUSSION

The results indicate that nine species of medium-large mammals occur in the Pomene National Reserve. These results differ from those of Macandza et. al. (2015), when studying the ecological and socio-economic conditions of the Pomene National Reserve, where they confirmed the presence of seven species of medium-large mammals, which were also found in this study. In comparison with previous similar studies carried out in the area, two species were added, the red kid (Cephalophus natalensis) and xipene (Raphicerus campestris). The open miombo was the habitat with the highest diversity of medium-large mammals in the Pomene National Reserve, with a Simpson's index of 0.846. These results can be supported by Roque (2015) who, when studying the assessment of faunal biodiversity in different levels of forest cover in the province of Manica, also found the greatest diversity of medium-large mammals in districts predominated by open miombo.The reason why the open miombo is more diverse than the dense miombo and the shrub prairie may be due to the fact that this habitat has a sufficient grass layer for foraging and trees for shade (Dunnin et. al., 1992). The open miombo has more scattered trees, and the grass layer is more abundant (PEA, 2016).The results of this study indicate that the African wild pig (Potamochoerus porcus) and the grey goat (Sylvicapra grimmia) were the most dominant species. Similarly, Roque (2015), when studying the assessment of faunal biodiversity at different levels of forest cover in the province of Manica, found that the African wild pig (Potamochoerus porcus) and the grey goat (Sylvicapra grimmia) were the species that also showed the highest proportion of occurrence. These The results differ from those obtained by the International Foundation for Fauna Management (2013), when studying the co-management of the Gilé National Reserve and its periphery, where the African wild pig (Potamochoerus porcus) was the second least dominant species in the study, with the grey goat (Sylvicapra grimmia) being the most dominant species. The higher proportion of occurrence of the African wild pig (Potamochoerus porcus) and the grey goat (Sylvicapra grimmia) can be explained by the fact that they find their preferred habitats, associated with their trophic strategies (the African wild pig is omnivorous and generalist, and the grey goat is a frugivorous herbivore) (Parr et. al., 2014 cited by Roque, 2015). However, omnivores are species with a less restricted diet compared to carnivores and herbivores.The results indicate that the shrub grassland was the habitat with the highest abundance of medium-large mammals compared to the miombo (dense and open). These results are similar to those of Passoscordeiro (1999), who found a greater number of individuals of terrestrial mammal species in open sites with higher grass cover. The higher abundance recorded in the shrub grassland can be explained by the fact that this habitat has a higher occurrence of some abundant species compared to the other habitats studied. The African wild pig (Potamochoerus porcus) and the golden-necked hare (Lepus saxatilis) are among the dominant species in the RNP and were the species with the highest frequency of occurrence in the shrub grassland in relation to the miombo (dense and open), as this habitat has sufficient quantities of forage to maintain the daily activities of the wildlife. Most of the species found are herbivores and the shrub grassland of the Pomene National Reserve is mostly made up of shrubs and a grass cover of up to 3,100 kg/ha of average biomass (PEA, 2016). Dense miombo and shrub prairie were the habitats that showed the greatest specific similarity, with a Jaccard similarity index of 0.875. These results can be explained by the proximity in terms of conditions and resources that these habitats offer to the animals. The shrub prairie is mostly made up of shrubs and a grass cover that provides a source of food for the animals, but there are no water sources in this habitat (PEA, 2016). When the animals need water, they move to the dense miombo, as this habitat has larger bodies of water, sufficient to satisfy the animals' need for water. During the day, the animals stay in the shrub grassland to forage. The results indicate that the grey goat (Sylvicapra grimmia), golden-necked hare (Lepus saxatilis) and African wild pig (Potamochoerus porcus) are the most frequently occurring species in the

Pomene National Reserve. These results are similar to those of Macandza et. al., (2015), when studying the ecological and socio-economic conditions of the Pomene National Reserve, where they found that the grey goat (Sylvicapra grimmia), golden-necked hare (Lepus saxatilis) and African wild pig (Potamochoerus porcus) were also the most widely distributed species in the Pomene National Reserve.The higher frequency of occurrence of these species can be explained by the fact that they are generalists and can occur in almost all types of terrestrial habitats (Parr et. al., 2014 cited by Roque, 2015).

# 6. CONCLUSION

According to the results obtained in this study, it can be concluded that:

➢ The Pomene National Reserve is home to nine species of medium-large mammals.

➢ The open miombo was the habitat with the greatest diversity of medium-large mammals.

➢ The African wild pig (Potamochoerus porcus) and the grey goat (Sylvicapra grimmia) were the most dominant species.

➢ In Pomene National Reserve, shrub grassland was the habitat with the highest abundance of medium-large mammals and open miombo the habitat with the lowest abundance.

➢ Dense miombo and shrub grassland were the habitats that showed the greatest specific similarity for medium-large mammals.

➢ The open miombo and the shrub grassland were the habitats with the highest occurrence of medium-large mammal species in Pomene National Reserve, with eight species in the open miombo and the same number in the shrub grassland.

➢ The occurrence of the chengane goat (Neotragus moschatus) was only recorded in the open miombo. The black-faced monkey (Cercopithecus pygerythrus) only occurs in the dense miombo and in the shrub grassland. The xipene (Raphicerus campestris) does not occur in the dense loam.

# 7. RECOMMENDATIONS

According to the results of this study, the Pomene National Reserve is recommended:

➢ The application of grazing management strategies, prioritising shrubby grassland and dense miombo as these are the habitats with the greatest abundance of herbivores;

➢ Carrying out studies on the habitat selection and diet of the species that occur in the Pomene National Reserve, in order to provide information on the habitats favoured by wildlife.

➢ Adjustment of the Reserve's boundaries.

# 8. BIBLIOGRAPHICAL REFERENCES


1. ANDRADE, A. H. (2017). Notions of sampling.

2. ARAUJO, A. C., & CHIARELLO, A. G. (2007). Camera traps in mammal sampling: methodological considerations and equipment comparison.

3. ARCHIE, E. A. & CHIYO, P. I. (2012). Elephant behaviour and conservation: social relationships, the effects of poaching, and genetic tools for management. Molecular Ecology.

4. BATISTA, V. G. (2012). Effects of habitat complexity, predation and proximity to forest fragments on the structure of tadpole assemblages in cerrado remnants in southern Brazil. 7pp.

5. BARROS, R. S. M. (2007). Measures of Biological Diversity: Postgraduate Programme in Ecology Applied to the Management and Conservation of Natural Resources - PGECOL. 2-8pp.

6. BARROSO, I., & POMBO, M. (2014). Influence of environmental factors on animal behaviour.

7. ,BARBETTA, P. A. (2002). Statistics Applied to the Social Sciences. Ed. UFSC, 5th Edition.

8. BEGON, M. (2006). Ecology from individuals to ecosystems. School of Biological Sciences.

9. BERNARDI, V. E., & LANDIM, P. M. B. (2002). Application of the Global Positioning System (Gps) in Data Collection. 4-11pp.

10. BERNARDO, P. V. S. (2012). Distribution patterns of medium and large mammals in fragmented landscapes. 27pp.

11. BIONDI, D. & BOBROWSKI, R. (2014). Using ecological indices to analyse the landscape treatment of trees in Curitiba's urban parks.

12. BOTHMA, J. P. (2002). Couting wild animals.

13. BROWER, J. E., & ZAR, J. H. (1984). Field & laboratory methods for general ecology.

14. CASSINI, S. T. (2005). Ecology: Fundamental concepts. 12pp.

15. CLAVETE, P. A. (2014). Spatial dynamics of elephant herd interaction in the Moribane Forest Reserve - Manica Province. 10pp.

16. CUNHA, F. P. (2013). Monitoring medium and large terrestrial mammals.

17. CUNHA, L. S. (2017). Types of sampling.

18. DCS - Delcam Consultoria e Serviços Lda (2016). Magoe National Park Management Plan.

19. DUNNIN, J. B., DANIELSON, B. J., & PULLIAM, H. R. (1992). Ecological processes that affect populations in complex landscapes.

20. EDUARDO M, (2014). Mammals.

21. ESTEVES, C. F. (2010). Anthropic Influence on the Spatial Distribution of the Mammal Community in Anchieta Island State Park. 43- 46pp.

22. FERNANDES, R. S., LITULO, C., LOURO, C. M. M., PEREIRA, M. A. M., & PEREIRA, T. I. F. C. (2017). Research and monitoring of species and ecosystems in marine conservation areas in Mozambique: Survey of Priorities and Capacities for the Implementation of Monitoring and Research Programmes. 4-5pp.

23. FILHO, J. L., & IOCHPE C. (1996). Introduction to Geographic Information Systems with emphasis on Databases. 2pp.

24. FRANCO, J. L. A. (2013). The concept of biodiversity and the history of conservation biology: from wilderness preservation to biodiversity conservation.

25. GEHRING, T.M., & SWIHART, R.K. (2013). Body size, niche breadth, and ecologically scaled responses to habitat fragmentation: mammalian predators in an agricultural landscape.

26. GOMES. A. S. & FERREIRA, S. P. (2004).Analysing of Analyses Ecological. Fluminense Federal University.

27. GOMES, T. S. (2010). GPS Fundamentals: Concepts, Operation and Configuration. 3pp.

28. HENRIQUES, S. (2012). Sampling.

29. HULLE, N. L. (2006). Medium and large mammals in a cerrado remnant in south-eastern Brazil. 50pp.

30. INIA. (1995). National Soil Chart, 1:1 m. INIA Land and Water Series. Maputo, Mozambique.

31. JOIA, D. G., MUCHANGA, M. V., & MAGODO, Z. J. (2016). Analysing The Problem of Uncontrolled Burning on Biodiversity in Conservation Areas - Gilé National Reserve (RNG).

32. LIMA, R. N. S. (2012). Google Earth applied to research and teaching of geomorphology. 17-22pp.

33. MACANDZA, V. OWEN-SMITH, A. N. & CROSS, P.C. (2004). Forage Selection by African buffalo in the Late Dry Season in Two Landscapes.

34. MACANDZA, V., et. al. (2015). Study of the Ecological and Socioeconomic Conditions of the Pomene National Reserve - Final Report: Project for the Sustainable Financing of Mozambique's Protected Areas System, Ministry of Land, Environment and Rural Development (MITADER), National Administration of Conservation Areas (ANAC).

35. MAE - Ministry of State Administration (2014). Profile of the district of Massinga, Province of Inhambane.

36. MAGUIRE,D. J., GOODCHILD,M.F.,&RHIND, D.(1991). Geographical Information Systems: Principles and Applications.

37. MANTOVANI, W. (1987). Floristic and phytosociological study of the herbaceous-subshrub layer in the biological reserve of Mogi Guaçu and Itirapina.

38. MARQUES, R. M., & RAMOS, F. V. (2001). Identification of mammals

occurring in the São Francisco de Paula National Forest/Ibama RS using photographic equipment triggered by infrared sensors.

39. MARTINI, A. M. Z., & PRADO, P. I. K. L. (2010). Indices of species diversity. 16pp.

40. MICOA. (2013). Environmental Profile and Mapping of Current Land Use in Mozambique's Coastal Zone Districts - Massinga District, Strategic Environmental Assessment of Mozambique's Coastal Zone. Maputo.

41. MITADER. (2015). Strategy and action plan for the conservation of biological diversity in Mozambique (2015-2035). 17pp.

42. MÜLLER-DOMBOIS, D.; ELLEMBERG, H. (1974). Aims and methods for vegetation ecology. 547p.

43. WORLD EDUCATION. (2014). Mammals.

44. PEA - Projectos e Estudos Ambientais (2016). Pomene National Reserve Management Plan.

45. NETO, P., VALENTINE, J., & FERNANDEZ, F. (1995). Topics in Biological data processing.

46. ODUM, E.P., BARRETT, G.W. (2008). Fundamentals of Ecology.

47. PASSOSCORDEIRO, J. L. (1999). Habitat classes and potential distribution of small terrestrial mammals (Rodentia, Sigmodontinae and Didelphimorphia) in the savannas of the middle and upper Surumu. 40pp.

48. PEACOR S. D., & WERNER E. E. (2001). The contribution of trait-mediated indirect effects to the net effects of a predator.

49. PEREIRA, V., & NAZERALI, S. (2016). Occurrence of Mozambique's threatened species in Parks, Reserves and Coutadas in 2016.

50. PERONI, N., & HERNÁNDEZ M. I. M. (2011). Ecology of Populations and Communities. 26-28pp.

51. PILLAR, V.D. (2002). Ecosystems, communities and populations: basic concepts. 3-4pp.

52. PRIMACK, R. (2006). Essentials of Conservation Biology.

53. REAL, R. (1999). Tables of significant values of Jaccard's index of similarity.

54. RODRIGUES, P. M., & REIS S. C. (2014). GPS system.

55. RODRIGUES, W. C. (2015). DivEs - Species Diversity v3.0 - User's Guide.

56. ROOS, F. L. (2010). The Use of Linear Transects for Monitoring Tree-Dwelling Mastofauna in the Mamirauá Sustainable Development Reserve.

57. ROQUE, D. V. & MACANDZA, V. (2015) Evaluation of Faunal Biodiversity in Different Levels of Forest Cover in the Province of Manica. 17pp.

58. Schneider, M. F., Buramuge, V. A., Luís A., & Serfontein F. (2005). Checklist and Centres of Vertebrate Diversity in Mozambique. 8- 10pp.

59. SEED - Sociedade de Engenharia e Desenvolvimento, (2003). Environmental Impact Study for the Cabo Delgado Biodiversity Tourism Project - CBDTP. Ministry for the Coordination of Environmental Action. 60pp.

60. SCHIESARI, L. (2016). Predation and community structure. 2pp.

61. SILVESTRE, R. (2009). Comparison of Floristics, Structure and Spatial Pattern in Three Fragments of Mixed Ombrophilous Forest in the State of Paraná. Master's thesis. Federal University of Paraná, Curitiba.

62. Simões, L. G. (2009). Determinants of mammal diversity and abundance in a Mediterranean agro-sylvo-pastoral system. 4pp

63. TOUMISTO, H. (2010). A Diversity of beta diversities, straightening up a concept gone awry.

64. VALERI, S. V., & SENÔ, M. A . A. F. (2003). The importance of ecological corridors for wildlife and the sustainability of forest remnants. 9pp.

65. IEIRA, M. T. & BESSEGATO, L. F. (2013). Notions of sampling.

66. WILD, H., & BARBOSA, L.A. (1967). Vegetation map of the Flora Zambesiaca area: Flora Zambesiaca Supplement. 71pp.

67. WITTEMYER, G., GETZ, W. M., VOLLRATH, F. & HAMILTON, D. I.(2007). Social dominance, seasonal movements, and spatial segregation in African elephants: a contribution to conservation behaviour. Behav Ecol Sociobiol 13.

# ANNEXES

## Annex 1: Mammal Diversity and Distribution Data Collection Form

### To//2018

### Habitat name:

### Transect n° :

| ID | Species | | Number of individuals | Location | |
|---|---|---|---|---|---|
| | Scientific name | Common name | | Latitude (S) | Longitude (E) |
| | | | | | |
| | | | | | |
| | | | | | |
| | | | | | |
| | | | | | |
| | | | | | |
| | | | | | |
| | | | | | |
| | | | | | |
| | | | | | |
| | | | | | |
| | | | | | |

Nelson Honiasse Pulaze Luis.

**Diversity and distribution of medium-large mammals in the Pomene National Reserve.**

# Annex 2: List of medium-large mammals i n   Pomene National Reserve

| ID | Scientific name | Common name | Family | Order |
|---|---|---|---|---|
| 1 | Lepus saxatilis | Golden-necked hare | Leoporidae | Lagomorpha |
| 2 | Sylvicapra grimmia | Grey goat | Bovidae | Artiodactyla |
| 3 | Cephalophus natalensis | Red kid | Bovidae | Artiodactyla |
| 4 | Papio ursinus | Grey dog monkey | Cercopithecidae | Primates |
| 5 | Cercopithecus albogularis | Simango monkey | Cercopithecidae | Primates |
| 6 | Potamochoerus porcus | African wild pig | Suidae | Artiodactyla |
| 7 | Raphicerus campestris | Xipene | Bovidae | Artiodactyla |
| 8 | Cercopithecus pygerythrus | Black-faced monkey | Cercopithecidae | Primates |
| 9 | Neotragus moschatus | Chengane kid | Bovidae | Artiodactyla |

# Annex 3: Distribution of medium-large mammals in Pomene National Reserve

| ID | Species | Dense miombo | Open miombo | Shrub prairie |
|---|---|---|---|---|
| 1 | Golden-necked hare (Lepus saxatilis) | It happens | It happens | It happens |
| 2 | Grey goat (Sylvicapra grimmia) | It happens | It happens | It happens |
| 3 | Red kid (Cephalophus natalensis | It happens | It happens | It happens |
| 4 | Grey dog monkey (Papio ursinus) | It happens | It happens | It happens |
| 5 | Simango monkey (Cercopithecus albogularis) | It happens | It happens | It happens |
| 6 | African wild pig (Potamochoerus porcus) | It happens | It happens | It happens |
| 7 | Xipene (Raphicerus campestris) | It doesn't happen | It happens | It happens |
| 8 | Black-faced monkey (Cercopithecus pygerythrus) | It happens | It doesn't happen | It happens |
| 9 | Chengane goat (Neotragus moschatus) | It doesn't happen | It happens | It doesn't happen |
| Number of species in each habitat | | 7 | 8 | 8 |

# Annex 4: Diversity, richness and abundance of medium-large mammals.

| Habitat | Simpson diversity index | Margalef Wealth Index | Abundance (%) |
| --- | --- | --- | --- |
| Dense miombo | 0,819 | 1,1007 | 33,0966 |
| Open miombo | 0,846 | 1,2967 | 31,3920 |
| Shrub prairie | 0,805 | 1,2678 | 35,5114 |

# Annex 5: Specific similarity between habitats.

| Habitats in comparison | Specific similarity |
|---|---|
| Dense Miombo and Open Miombo | 0,667 |
| Open Miombo and Shrub Prairie | 0,875 |
| Dense Miombo with Shrubby Grassland | 0,778 |

# Annex 6: Determination of diversity e dominance of medium-large mammals in the three habitats of the Reserva Nacional de

| Dense miombo | ni | Dominance (%) | ni*(ni-1) | I.simpson = $1 - \dfrac{\sum_{i=1}^{n} ni(ni-1)}{N(N-1)}$ |
|---|---|---|---|---|
| Red kid (Cephalophus natalensis) | 44 | 18,8841 | 1892 | 0,81911 |
| Grey goat (Sylvicapra grimmia) | 58 | 24,8927 | 3306 | |
| Golden-necked hare (Lepus saxatilis) | 21 | 9,0129 | 420 | |
| African wild pig (Potamochoerus porcus) | 56 | 24,0343 | 3080 | |
| Black-faced monkey (Cercopithecus pygerythrus) | 9 | 3,8627 | 72 | |
| Grey dog monkey (Papio ursinus) | 27 | 11,588 | 702 | |
| Simango monkey (Cercopithecus albogularis) | 18 | 7,7253 | 306 | |
| Total | 233 | 100 | 9778 | |
| Open miombo | | | | |
| Golden-necked hare (Lepus saxatilis) | 26 | 11,7647 | 650 | 0,84603 |
| Grey goat (Sylvicapra grimmia) | 45 | 20,362 | 1980 | |
| Red kid (Cephalophus natalensis) | 22 | 9,9547 | 462 | |
| Grey dog monkey (Papio ursinus) | 39 | 17,6471 | 1482 | |
| Simango monkey (Cercopithecus albogularis) | 33 | 14,9321 | 1056 | |
| African wild pig (Potamochoerus porcus) | 42 | 19,0045 | 1722 | |
| Chengane goat (Neotragus moschatus) | 12 | 5,4299 | 132 | |
| Xipene (Raphicerus campestris) | 2 | 0,905 | 2 | |
| Total | 221 | 100 | 7486 | |
| Shrub prairie | | | | |
| Grey goat (Sylvicapra grimmia) | 57 | 22,8 | 3192 | 0,80498 |
| African wild pig (Potamochoerus porcus) | 69 | 27,6 | 4692 | |
| Grey dog monkey (Papio ursinus) | 31 | 12,4 | 930 | |
| Red kid (Cephalophus natalensis) | 19 | 7,6 | 342 | |
| Xipene (Raphicerus campestris) | 5 | 2 | 20 | |
| Black-faced monkey (Cercopithecus pygerythrus) | 9 | 3,6 | 72 | |
| Golden-necked hare (Lepus saxatilis) | 54 | 21,6 | 2862 | |
| Simango monkey (Cercopithecus albogularis) | 6 | 2,4 | 30 | |
| Total | 250 | 100 | 12140 | |

**Pomene**

## I want morebooks!

Buy your books fast and straightforward online - at one of world's fastest growing online book stores! Environmentally sound due to Print-on-Demand technologies.

Buy your books online at
**www.morebooks.shop**

Kaufen Sie Ihre Bücher schnell und unkompliziert online – auf einer der am schnellsten wachsenden Buchhandelsplattformen weltweit! Dank Print-On-Demand umwelt- und ressourcenschonend produzi ert.

Bücher schneller online kaufen
**www.morebooks.shop**

Printed by Books on Demand GmbH, Norderstedt / Germany